Taculator App Guidebook

Your Tutor to Learn How the Taculator Graphing Calculator App Works with Screenshots & Keystroke Sequences

Marco Wenisch

Copyright © 2020 by Marco Wenisch. All Rights Reserved.

No part of this publication may be reproduced, distributed, or transmitted in any form or by any means, including photocopying, recording, or other electronic or mechanical methods, or by any information storage and retrieval system without the prior written permission of the publisher, except in the case of very brief quotations embodied in critical reviews and certain other non-commercial uses permitted by copyright law.

For permissions contact:
marco@marcowenisch.com

TABLE OF CONTENTS

1 THE TACULATOR APP 1

   1.1   DOWNLOAD THE APP 1
   1.2   INITIAL COLOR SCHEME SETUP 1
   1.3   TACULATOR PRO 2
   1.4   GETTING HELP & CONTACT 4

2 FIRST STEPS 5

   2.1   THE BASICS OF THE BASICS 5
      A. USING THE SHIFT KEY 5
      B. USING THE ALPHA KEY 5
      C. DELETE AND EDIT 6
      D. EDIT PREVIOUS ENTRIES 6
      E. UNDO & REDO 6
      F. SELECTING MENU COMMANDS 6
      G. COMMANDS CATALOG 7
   2.2   MODE SETTINGS 8
   2.3   BASIC ARITHMETIC 10
      A. NEGATION KEY 10
      B. ENTER EXPONENTS 11
      C. ENTER ROOTS 11
      D. USING PARENTHESES 11
      E. PREVIOUS ANSWER 12
   2.4   STORING VARIABLES 12
      A. STORE VARIABLES 12
      B. RECALL VARIABLES 13
   2.5   WORKING WITH FRACTIONS 13
      A. ENTER FRACTIONS 13
      B. CONVERTING FRACTIONS 14
   2.6   TESTING NUMBERS 15
   2.7   CONVERTING ANGLES & DMS 16
      A. DEGREE TO RADIAN 16
      B. RADIAN TO DEGREE 16
      C. DEGREE TO DMS 17
      D. OVERRIDE MODE OF ANGLES 17

  E. Entering Angles in DMS 17
 2.8 Sharing Equations with Friends 18

3 GRAPHING BASICS 19

 3.1 Enter Functions 19
  A. Entering Functions 19
  B. Deselect Functions 20
  C. Families of Functions 20
 3.2 Formatting the Graph 21
 3.3 Setting the Graph Window 22
 3.4 Zooming the Graph Window 23
 3.5 The Graph Menu 25

4 DIFFERENTIAL CALCULUS/ ANALYZING FUNCTIONS 27

 4.1 Tracing a Graph 27
 4.2 Find Y-Value 28
 4.3 Find X-Value 29
 4.4 Y-Intercept 30
 4.5 Zeros of a Function 31
 4.6 Minimum 32
 4.7 Maximum 33
 4.8 Intersection of two Functions 34
 4.9 Draw Derivative 35
 4.10 Inflection Points 36

5 SOLVE EQUATIONS 37

 5.1 Polynomial 37
 5.2 Solve any Equation 38
 5.3 Solve Quadratic Equations 39

6 INTEGRAL CALCULUS 40

 6.1 Calculate Integral 40
 6.2 Integral in GRAPH Menu 41
 6.3 Find Area with Absolute Value 42
 6.4 Area between two Functions 44
 6.5 Integral Function 45

7 MATRICES 46

7.1 SAVE MATRIX 46
7.2 DELETE MATRIX 48
7.3 PUT INTO ROW-ECHELON FORM 49
7.4 SOLVE MATRIX (REDUCED ROW-ECHELON FORM) 50
7.5 TRANSPOSE A MATRIX 51
7.6 IDENTITY MATRIX 52
7.7 INVERSION OF A MATRIX 53
7.8 DETERMINANT OF A MATRIX 54
7.9 MATRIX ARITHMETIC OPERATIONS 55

8 STATISTICS & PROBABILITY 56

8.1 PERMUTATIONS, COMBINATIONS & FACTORIALS 56
A. FACTORIALS 56
B. PERMUTATION AND COMBINATION 56
C. BINOMIAL THEOREM 57
8.2 RANDOM NUMBERS 58
A. RANDOM DECIMALS 58
B. RANDOM INTEGERS 58
C. RANDOM INTEGERS NO REPETITION 59
8.3 LISTS & STATISTICAL DATA 60
A. ENTER AND DELETE LISTS 60
B. INSERT A NEW LIST 61
C. SORTING LISTS 61
8.4 HISTOGRAM, BOX PLOT & SCATTER PLOT 62
A. HISTOGRAM AND BOX PLOT 62
B. ADJUST THE CLASS SIZE 63
C. TWO-VARIABLE DATA PLOTS 64
D. TRACING PLOTS 66
8.5 STATISTICAL DATA ANALYSIS 66
A. ONE-VARIABLE DATA ANALYSIS 66
B. TWO-VARIABLE DATA ANALYSIS 67
C. MORE STATISTICS COMMANDS 68
8.6 REGRESSION 69

1 THE TACULATOR APP

1.1 Download the App

Taculator is available to download for iPhone and iPad on the App Store. Open the App Store and search for "Taculator". Look for the app "Taculator Graphing Calculator" and press the GET button. The app is free to download.

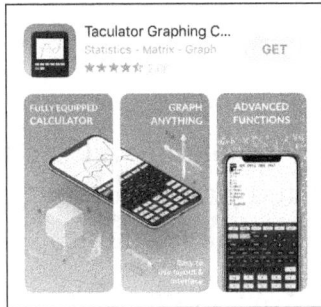

Alternatively, you can visit the website www.taculator.com that provides a link to download the app.

1.2 Initial Color Scheme Setup

Once you've installed the Taculator app and then open it, you will see the screen "Create your own calculator". Taculator is fully customizable and allows you to change the background color, as well as the color of the Shift and Alpha button, the light upper and lower keys, and the dark middle keys. Simply tap on the color of your choice and press the "Next" button at the bottom right to proceed.

 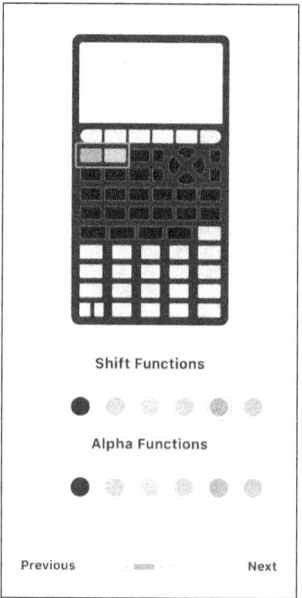

If you do not make any changes to the colors, Taculator will use the standard color scheme which is black/grey and red/orange for the Shift and Alpha key. In this book I'm using the standard color scheme.

Once you are done, press the "Finish" button. You can now start using the Taculator Graphing Calculator.

Later, you can change the colors of your calculator in the mode menu. Press the **mode** key and tap on "Change Keyboard Color".

1.3 Taculator PRO

The Taculator app is free to download and already offers a lot of functions for free. The complete scientific calculator together with commands like sin, cos, tan, ln, e, log, root, etc. are all available to use for free.

Furthermore, the graphing and tracing of functions can be done on the free version, too. It's also possible to share functions and equations with your contacts and to read through the full command catalog.

By using the free version of Taculator you can get a sense of how the app works and if you like it, you can upgrade to Taculator PRO to be able to use over 100 advanced calculator commands.

While graphing is available in the free version, investigating the functions for zero, min, max, intersect, dy/dx, and finding the integral is only available in the PRO version.

Additionally, most of the commands of the Math, Matrix, Test, Angle, Draw, List, Statistics, and Distribution menu require Taculator PRO.

To get Taculator PRO, click on any command that has a blue PRO label at the right. For example, go to the `math` menu and press fMin `6` to use this command. Because it is locked, the app will show you a screen to get Taculator PRO.

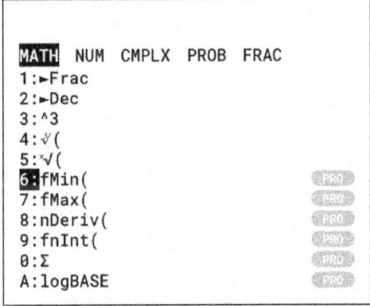

There are three options to choose from:

- A monthly subscription.
- A yearly subscription.
- A lifetime subscription which is a one-time payment.

You can cancel the monthly and yearly subscription anytime for any reason. It must be at least 24 hours before the current subscription period ends.

1.4 Getting Help & Contact

In case you need help with anything or having a problem, you can press the help ⟨?⟩ key located at the bottom left. It opens a menu where you can "Send a Suggestion" or "Report a Problem". Pressing one of those buttons opens the mail app to send an email to the Taculator Support.

This menu also provides a brief tutorial about how to use the core functionalities of the app as well as a button to change the keyboard color.

Another way to get help for the specific calculator commands it to open the catalog ▢.

2 FIRST STEPS

2.1 The Basics of the Basics

When you must press several keys in a row, keys are always pressed one at a time. Never press more than one key simultaneously.

a. Using the SHIFT key

Most keys have secondary key functions to access various calculator menus or to input mathematical symbols in an expression. These functions are written in a different color (depending on the color you've set up for your calculator keyboard). They are located on the top left of the button itself and can be used by pressing the [shift] key first.

Alternatively, you can press and hold a button that automatically triggers the second function. You do not have to press the [shift] key in this scenario. When the [shift] key is active, it gets a white-colored border and all the secondary functions are highlighted with their respective color.

b. Using the ALPHA key

Above most of the keys, a letter (A – Z) is written on the right having the color you have set up for your calculator keyboard. Those letters can be used to store a variable or to access shortcuts. Just press the [alpha] key to access those.

To enter many letters consecutively, you can lock into Alpha mode. Then your calculator will only enter the letters written above the keys. Press [shift] > [alpha] to activate the Alpha-Lock and press [alpha] again to exit the Alpha-Lock.

c. Delete and Edit

- To delete parts of your entry, use the arrow keys [arrows] to place the cursor right of the character you want to delete and press [delete].
- Taculator supports a touch-enabled interface, so alternatively from using the arrow keys, you can use your finger to tap on any position you want to place the cursor.
- If you want to erase everything you've entered, press the [clear] key.
- The calculator will place new characters left to the cursor and also delete the character left to the cursor.

d. Edit Previous Entries

To recall already entered commands or expressions, press the [↑] arrow key to scroll through all previous entries or use your finger to scroll up. Highlight the entry you want to use again and press [enter]. It will be pasted into your current entry line, where you can edit it and do the calculation again. Another way to do this is to highlight the entry, wait for the tooltip to appear, and then tap on COPY.

e. Undo & Redo

Besides using the [delete] key, you can also use the keys for undo [↶] and redo [↷]. The keys can be used while you are entering an equation on the home menu. The undo key will remove the last character that has been inserted into the calculator.

f. Selecting Menu Commands

Most functions are located inside different menus. They can either have their own button like MATH or STATS, or can be accessed by pressing the [shift] button first. All of the different menus have their own chapter in this book to explain them.

FIRST STEPS 7

Keep in mind that most menus have more than one page of commands. You can switch between pages with the arrow keys ▶ ◀, or use your finger.

To select menu items, one possibility is to highlight them by using the arrow keys ▲ ▼ (respectively your finger) and hitting [enter].

As you can see, the menu items are numbered 1 to 9, and then continued with A-Z. Pressing the appropriate key will also select and paste them into the calculator.

To quit any menu and return to the home menu, press [shift] > exit [mode].

g. Commands Catalog

Using some commands can be difficult as they require to use proper syntax in order to calculate it correctly. Taculator includes a full catalog of all commands with a syntax help, description, example, and possible errors. To open the catalog, press the catalog ▣ key. A new menu opens up and shows all the different menus with the respective commands inside. You can also use the search field to search for a specific command.

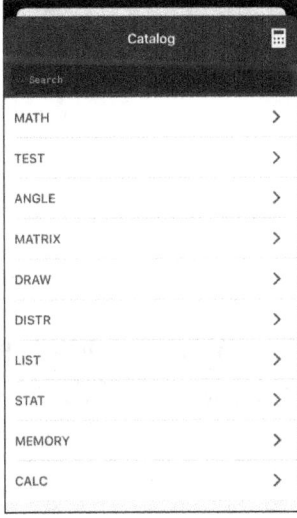 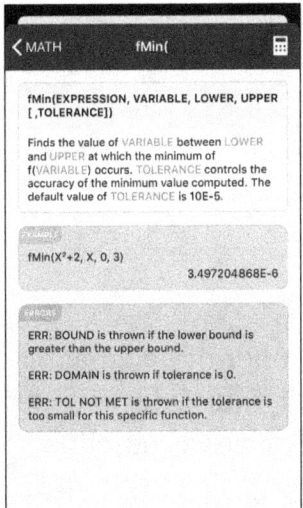

2.2 Mode Settings

Choosing the right mode settings is very important in getting the most out of Taculator. In this chapter, I will explain all settings in this menu.

Open the mode menu by pressing [mode].

Now you can change the settings by highlighting the item in each row. Use the ✪ arrow keys to navigate in the mode menu, or tap on the item you'd like to change with your finger and use [enter] to select it.

NORMAL SCI ENG:

This setting changes how numbers are displayed on the calculator. For **NORMAL**, numbers are displayed in their usual numeric fashion for up to ten digits. For larger numbers, the calculator will use the scientific mode (**SCI**) automatically. Engineering mode (**ENG**) is just another way to display numbers. I recommend using NORMAL mode.

This is how the calculator will display numbers for the different modes.

(NORMAL: 500000, SCI: 5e5, ENG:500e3)

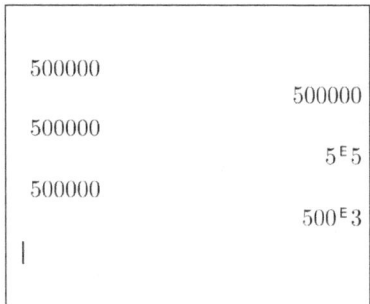

FLOAT 0123456789

Here you can set the number of decimal places to be rounded to. For example, selecting **2** will round to two decimal places. **FLOAT** will display as many decimal places as possible.

RADIAN DEGREE

This setting changes how angles are interpreted by the calculator. If you are working with trigonometric functions (such as sine, cosine, and tangent), you would choose **RADIAN**. Then the functions will be graphed for $-2\pi < x < 2\pi$. If you use **DEGREE**, $-360 < x < 360$ would be needed as the limits for the x-axis.

REAL a+bi re^(Oi)

This setting lets you choose between real and complex numbers. For complex numbers, you can use a+bi to display complex numbers in rectangular form and re^Oi to display them in polar form.

FONT1 FONT2 FONT3

In Taculator you can choose between three different fonts. The screenshots below show you how each font looks like (from left to right -> FONT 1 to 3). The differences are rather small, but you might have a preference. My favorite font is FONT1.

SIZE1 SIZE2 SIZE3

Like the font, you can choose between three sizes for the font. SIZE1 is the smallest and SIZE3 the biggest font. SIZE3 is best for large screens if you are using Taculator on an iPad for example. The screenshots below compare the different sizes.

CHANGE KEYBOARD COLOR

Clicking on this brings you directly to the screen "Build Your Calculator". Here you can customize the looks of your calculator. You can change the color of the background or of any button you would like. Just make sure that you still can see all the shift and alpha functions as it is possible to select the same color for those functions and the background.

2.3 Basic Arithmetic

If you start the app and it was closed before, it will always show you the home menu where you can do arithmetic operations. If you are in any other menu and want to go back, you can press the [on] button or press [shift] > exit [mode].

Basically, you can enter expressions in Taculator like you would write them on paper. However, there are some exceptions that you need to look out for:

a. Negation Key

If you want to enter a negative number, you must use the negation [(-)] key. It would be wrong to use the subtraction [−] key and can lead to incorrect calculations or error messages. There's a big difference between the two keys!

b. Enter Exponents

The [^] key with the caret symbol is used to enter exponents. Keep in mind that the cursor jumps up to the exponent position after you press this button and it will remain there until you press the right arrow ▶ key. Or you can tap right of the exponent with your finger to bring down the cursor. The keyboard of the calculator also includes the [x²] key to square a number with one click.

c. Enter Roots

To enter a square root, press [shift] > √[x²] and type the expression you would like to evaluate. The cursor remains under the radical sign as long as the right arrow ▶ key is not pressed.

The symbol for the **nth root** can be found in the math menu. Press [math] > ˣ√ [5] to paste the root symbol to the home menu. Then, enter the index number of the root, press the right arrow ▶ key, and type the rest of your expression.

d. Using Parentheses

You will need to use parentheses for some operations.

Furthermore, I highly recommend that you always surround negative numbers with parentheses if you want to raise them to a power. Let's say you want to square -2 and enter -2^2 in the calculator; you would expect to see 4 as the answer. However, your calculator will display -4. This is because the negation key [(-)] gets interpreted as -1 * 2 in our example. So the calculator would square 2, which is 4 and then multiply by -1 and the result will be -4. Always surround negative numbers with parentheses.

```
cos(5) + 3
                    3.283662185
cos(5 + 3)
                    -0.1455000338
-2²
                    -4
(-2)²
                    4
```

e. Previous Answer

If you start a new entry with an arithmetic operator, [+], [−], [×], [÷], or [^], the calculator will automatically insert **ans** which means that the previous answer will be used for the calculation. This won't happen if you start a new entry with numbers or functions. If you want to use a previous answer in your new entry, you can press [shift] > ans [(-)] to call the previous answer.

2.4 Storing Variables

Storing numbers or expressions can save time if you need them often. For example, if you need to use a number like "1.8634" several times, you can store it as the letter A, so A will be this number.

a. Store Variables

Enter the number or expression you want to store. Then press the store key [→] which will enter an arrow. Afterwards, enter the letter you want to store it to (press the [alpha] key and the key belonging to letter you'd like to use).

b. Recall Variables

If you forget what you have stored in a particular letter, you can enter the letter and press `enter`. The calculator will show its value.

However, if you are entering an expression at the moment and you need to look up a letter without aborting, you can use `shift` > rcl `→`. Then enter the letter you want to look up. Note that the calculator will insert it into your current entry.

2.5 Working with Fractions

To enter a fraction, simply press the `n/d` key. First enter the numerator and press the right arrow ◆ key to jump down to enter the denominator. Of course, you can always use the division `÷` key to create a fraction, too. However, this fraction won't look the way it would when you write it on paper.

a. Enter Fractions

Another way to enter fractions and to have some more choices is to open the FRAC shortcut menu by pressing `alpha` > F1 `y=`. A menu with four choices will pop up:

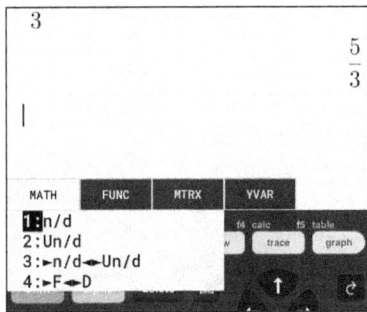

1. **n/d:** Pastes the fraction template to enter common fractions (usually you would use this option).
2. **Un/d:** Just like the first option but to enter mixed numbers and fractions.
3. **>n/d<>Un/d:** Converts a mixed number to a fraction or vice versa.
4. **>F<>D:** Converts decimals to fractions and vice versa.

Use the arrow keys ⬆ ⬇ to choose between the options and press [enter] to paste them.

Use the n/d fraction multiple times in the same fraction to enter complex fractions.

b. Converting Fractions

There are two functions to convert a fraction to a decimal (Dec) or to convert a decimal to a fraction (Frac). They are located in the MATH menu, so press [math] and insert Frac [1] or Dec [2]. Confirm with [enter].

If you have an infinite repeating decimal and want to turn it into a fraction, you can simply type the first ten digits of the number and use the Frac or F<>D command.

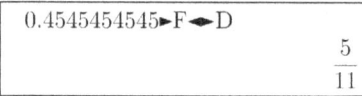

2.6 Testing Numbers

The functions of the TEST menu look very simple at first glance, but they can really help you compare complex expressions.

Enter the first expression and press [shift] > TEST [math] to access the TEST menu. You can choose between six relational operators by pressing the [1] through [6] keys. After that enter your second expression and press [enter].

The calculator's answer will be 1 or 0.

- 1 means TRUE
- 0 means FALSE

```
3 + 2 > 6
                    0
3 + 2 < 6
                    1
```

Your teacher could ask you on a test:

Evaluate ln(2) + ln(4)

1. ln(2-4)
2. ln(2+4)
3. ln(2*4)
4. ln(2/4)

If you don't know the answer, you can use your calculator and test all the answers. The answer 1 means that the statement is true.

```
ln(2) + ln(4) = ln(2 − 4)
                              Error
ln(2) + ln(4) = ln(2 + 4)
                              0
ln(2) + ln(4) = ln(2 * 4)
                              1
ln(2) + ln(4) = ln(2 ÷ 4)
                              0
|
```

2.7 Converting Angles & DMS

a. Degree to Radian

Make sure your calculator is in Radian mode, and you are on the home screen. First, enter the number of degrees you want to convert to radians. Then press [shift] > ANGLE [apps] to access the ANGLE menu. Finally, press [1] to paste in the ° symbol and press [enter].

If you like to see radian measures expressed as a multiple of π, do the following: Divide the radian measure by π and convert the result to a fraction by using the Frac command [math] > Frac [1].

```
60°
                    1.047197551
ans ÷ π
                    0.3333333333
ans ► Frac
                              1
                              ─
                              3
```

60° in degrees is 1/3π in radians.

b. Radian to Degree

Make sure your calculator is in Degree mode, and you are on the home screen. First, enter the radian measure you want to convert to degrees.

Surround the arithmetic expression with parentheses. Then press shift > ANGLE apps to access the ANGLE menu. Finally, press 3 to paste in the ʳ symbol and press enter.

c. Degree to DMS

Make sure your calculator is in Degree mode, and you are on the home screen. First, enter the degree measure. Then press shift > ANGLE apps to access the ANGLE menu. Finally, press 4 to paste in the DMS function and press enter.

d. Override Mode of Angles

You can easily force your calculator to use the angle units you want no matter what mode setting is active. To do so enter the number of degrees or radians and add the degree ° or radian r symbol. Press shift > ANGLE apps to access the ANGLE menu and press 1 or 3 to paste in the symbol. This will force your calculator to use the units you want regardless of the mode setting:

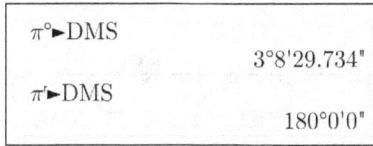

e. Entering Angles in DMS

Follow these steps to enter an angle in DMS measure:

1. Enter the degree measure and press shift > ANGLE apps > 1 to paste in the ° symbol.
2. Enter the minutes and press shift > ANGLE apps > 2 to paste in the ' symbol.
3. Enter the seconds and press alpha > " + to paste in the " symbol.

2.8 Sharing Equations with Friends

Another quite useful feature of Taculator is to share equations with your friends by using mail or any messenger app. You can share equations from the home menu as well as matrices, lists, and Y= functions. The person who receives it can click on the link and paste it into Taculator again on their device.

To share a function from the home menu, tap on the equation, and click the Share button that appears.

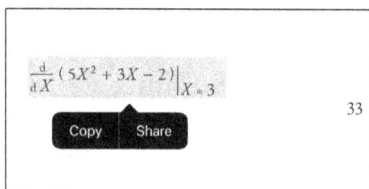

To share a Y= function, matrix, or list, you must press the ⬆ key to open the share menu. From here you can select what you want to share.

3 GRAPHING BASICS

3.1 Enter Functions

a. Entering Functions

In order to graph functions, you have to enter them. Press `y=` to access the Y= editor and enter your first function. After you are finished, press `enter` which will bring the cursor to the next line. You can enter another function here or press `graph` to graph the function.

Remember that the calculator only allows the letter X, inserted by the `x` key for the independent variable.

You can also use an already stored function when entering new functions. For example, you want to subtract one function from another. Just enter both functions as Y1 and Y2 and enter "Y1-Y2" as function Y3. That makes it clearer and can save time.

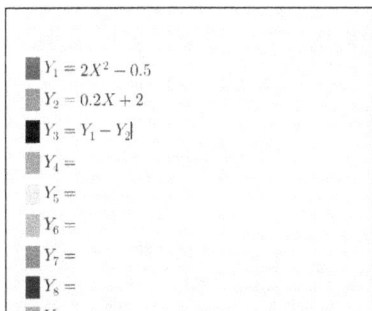

To paste a function name (Y1, Y2, ...), press `alpha` > F4 `draw` to open the shortcut menu. Use the arrow keys to select the different functions and press `enter` to paste them in.

b. Deselect Functions

If you deal with multiple functions, you may want to draw only one function at a time, while the others are still stored. That's absolutely possible and quite easy. To deselect a function, go to [y=] and use the ✪ arrow keys to place the cursor in the line of the equation you want to deselect and press [enter]. The function gets grayed out which means that it won't be graphed. To turn it back on, place the cursor in the line of the function and press [enter] again.

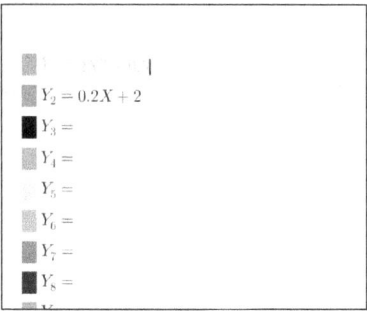

c. Families of Functions

In case you want to graph a family of functions, enter all numbers for the parameter inside brackets {} separated by commas.

The example below demonstrates the procedure to enter the function f(x)=ax.

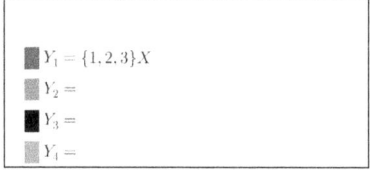

Putting the numbers {1,2,3} inside brackets means the calculator will draw 1X, 2X, and 3X.

3.2 Formatting the Graph

The settings made in the FORMAT menu will affect the appearance of the GRAPH menu.

Open the FORMAT menu by pressing [shift] > FORMAT [zoom].

Now you can change the settings by highlighting an item in each row. Use the ✪ arrow keys or your finger to navigate and use [enter] to select menu items.

The best way to go is to select all the options on the left. If you want to know more, read the following section, which explains all the format settings.

CoordOn CoordOff

ON means the calculator will show the coordinates of the cursor at the bottom of the screen. If you turn this off, you won't see the coordinates anymore if you trace a function.

GridOn GridOff

This setting turns the grid lines on or off. I find it quite helpful to have this set to GridOn. The screenshot below shows the difference of GridOn and GridOff.

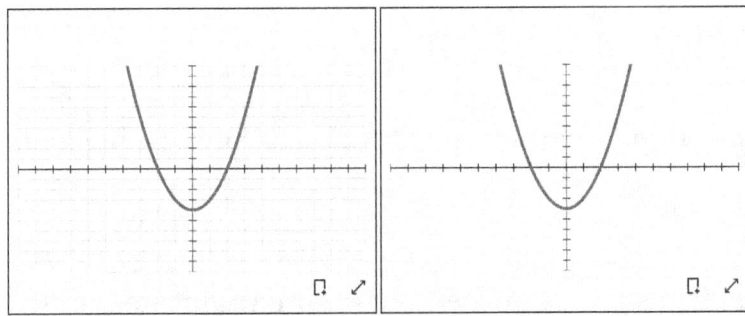

AxesOn AxesOff

Like for the grid, you can decide to turn on or off the coordinate axes. I don't see a reason to turn off the axes, so I recommend setting it to AxesOn.

3.3 Setting the Graph Window

Your calculator is limited to the x- and y- coordinates set in the WINDOW menu when it graphs a function. Therefore, you should know how to properly set the data in this menu.

Open the WINDOW menu by pressing [window].

Use the ✥ arrow keys to navigate and simply override existing values by entering new numbers, or press [clear] and enter a new number.

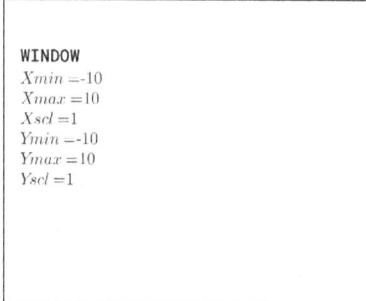

Xmin & Xmax

Sets the start and end value of the x-axis. Make sure Xmin is smaller than Xmax.

Xscl

You can change the distance between the tick marks located on the x-axis. The smaller the number, the smaller the distance between them and the more tick marks will appear.

Ymin & Ymax

Sets the start and end value of the y-axis. Make sure Ymin is smaller than Ymax.

Yscl

You can change the distance between the tick marks located on the y-axis. Remember that the number to enter here must be greater than 0 as the calculator won't show any tick marks for negative numbers.

3.4 Zooming the Graph Window

Using the ZOOM functions of your calculator makes it easier to adjust your graph window than setting the coordinates in the WINDOW menu. You can choose between several commands which will be explained in this chapter.

Open the ZOOM menu by pressing [zoom]. Notice that you don't have to graph your functions before using ZOOM commands. You can access the ZOOM menu directly from the Y= menu, just select a command and it will subsequently be graphed.

ZDecimal

This command zooms the window to the following coordinates: $-6.6 \leq x \leq 6.6$ and $-4.1 \leq y \leq 4.1$.

ZSquare

Your current window will be readjusted to graph elements like a circle properly. Setting the x and y coordinates to the same value would distort the circle as the calculator screen has a larger width than height.

GRAPHING BASICS

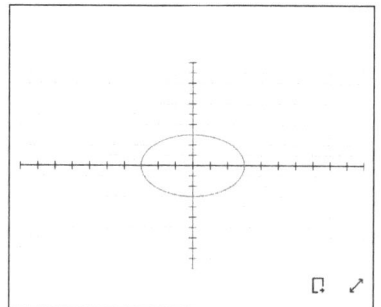
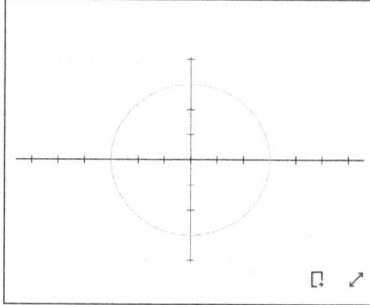

To draw a circle, start from the main menu and press the [draw] key to access the DRAW menu. Select Circle [7] to paste the command to the main menu. Then enter your X- and Y-value (the center point of the circle), separated by a comma and specify the radius of the circle. Press [enter] to draw the circle. For example: Circle(0,0,3). This will draw a circle with a radius of 3 and a center point of (0|0). To delete the circle again, press [draw] > ClrDraw [1].

ZStandard

This command zooms the window to the following coordinates: $-10 \leq x \leq 10$ and $-10 \leq y \leq 10$. If you don't know the dimensions of your function, ZStandard is always a good option to start with to get a sense of how it looks like.

ZTrig

Use this command to set the window when graphing trigonometric functions. It provides the best settings for such functions, like setting the tick marks on the x-axis to $\pi/2$.

ZoomStat

Only use this command to find a suitable viewing window if you are graphing plots. For normal functions, it will be useless.

ZoomFit

This command automatically zooms the y-coordinates of the window to fit your graph well. It doesn't affect the x-axis.

3.5 The Graph Menu

The previous chapters already described how to enter functions and set up the right format and window settings, as well as how to zoom the graph. This chapter tells you some more about what you can do with your graphs.

If you just want to zoom in and out to take a closer look at the graph, you can use your fingers. Double-tap to zoom in evenly. However, this way the zoom window will be reset once you want to trace the function or find points like zeros of the equation. To keep the zoom settings you'd like, you have to use one of the commands described above or set the window settings yourself.

Taculator lets you view the graph in full screen mode. To do this, tap on the little arrow ↗ icon right under the graphing window. In full screen mode, zooming and tracing is enabled.

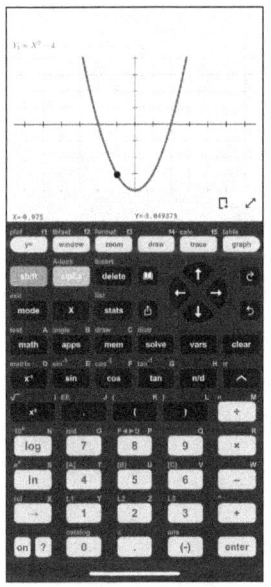
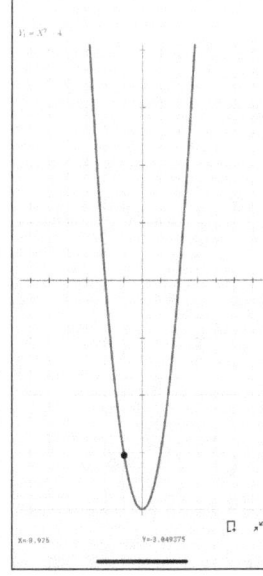

Another thing you can do is to share your graphs. Press the share icon ⌴ which is located right under the graphing window and select "Image" or "PDF". You can save the graph file to your phone or directly send it to your friends. You can do this in normal mode or full screen mode which will change the size of the graphing window.

4 DIFFERENTIAL CALCULUS/ ANALYZING FUNCTIONS

4.1 Tracing a Graph

To trace a function, press `trace` and you will see that a point on the function appears (indicating the position of the cursor), together with the X- and Y- coordinates below the graph and the equation of the function on top of the graph. Now you can use the right or left arrow keys ◆ ◆ to investigate the function. Taculator always starts tracing the function stored in Y1. To trace another function, use the up and down arrow keys ◆ ◆.

Even easier is to just use your finger to trace the function. You can also do this without pressing the `trace` key.

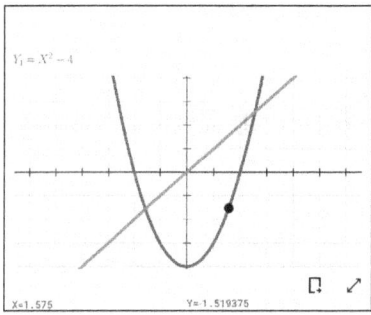

While you are tracing the graph, you can enter any x-value and the cursor will jump to that point. Make sure the value is between your window settings for Xmin and Xmax (so that you can see the point in your graphing window).

You can change the amount the x-value changes each time you move the cursor right or left. Press `window` and move your cursor all the way down to the ΔX line. Override the value with your desired trace step value.

4.2 Find Y-Value

Press shift > CALC trace to access the CALC menu. Select value 1 and enter the x-value of the point where you want to find the y-value and press enter.

If the display of the calculator is not showing your entered x-value, it will still show you the correct y-value but you won't see a point on the graph.

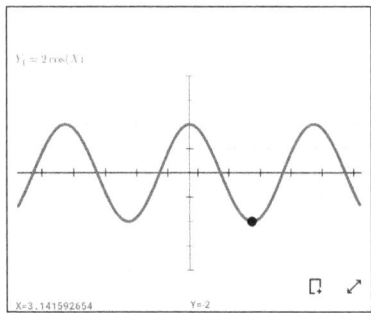

The screenshot shows the y-value of the function Y1=2cos(X) at the point X=π.

4.3 Find X-Value

There is no function to directly enter a y-value and find the x-value of the function at that point. You have to enter your y-value as a new function and find the intersection of both functions. This will show you the x-value.

First enter your y-value as a new function in the [y=] menu. Then press [shift] > CALC [trace] to access the CALC menu. Press intersect [5], select both functions by pressing [enter] and make a guess by entering an approximate x-value.

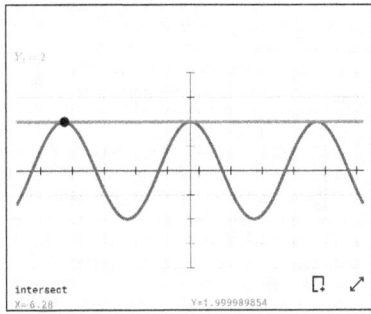

Finding one x-value of the function Y1=2cos(X) at the point Y=2.

4.4 Y-Intercept

A quick and easy way to find the intersection with the y-axis (x=0) is to paste the name of your function into the home menu. Go to the Y= editor and enter your function, then follow the steps below to find the y-intercept.

Make sure you are in the home menu and press **vars** > Functions **1** > select and enter the Y- function you want to use. Insert a parenthesis **(** followed by the x-value which is 0 for the y-intercept and close the parenthesis **)**. Press **enter** to execute the command.

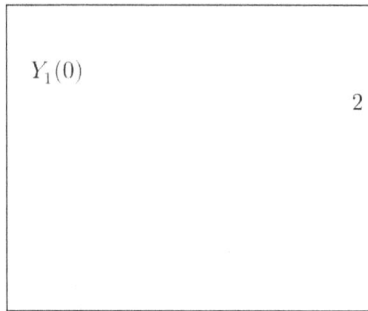

Finding the y-intercept of Y1=2cos(X).

4.5 Zeros of a Function

The zeros of the functions are the points where the graph of the function intersects or touches the x-axis. In addition, they are the solutions of the equation f(x)=0.

Press **shift** > CALC **trace** to access the CALC menu. Select zero **2** and set the left bound by tapping on the graph with your finger to place the cursor left of the zero. Alternatively, you can enter an x-value and press **enter**. Do the same for the right bound to enclose the zero point but make sure that the bounds include one zero of the function only.

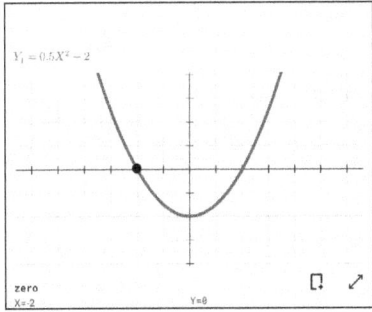

One zero of the function Y1=0.5X²-2 at X=-2.

4.6 Minimum

Press [shift] > CALC [trace] to access the CALC menu. Select minimum [3] and set the left bound by tapping on the graph with your finger to place the cursor left of the minimum. Alternatively, you can enter an x-value and press [enter]. Do the same for the right bound to enclose the minimum but make sure that the bounds include one minimum of the function only.

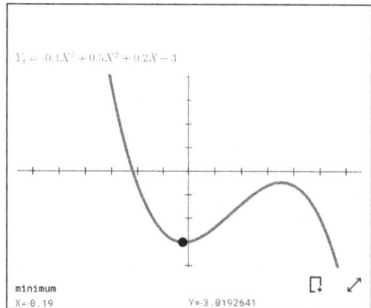

Finding a minimum (turning point) at (-0.19|-3.019) of the function Y1=-0.1X³+0.5X²+0.2X-3.

4.7 Maximum

The way to find a maximum is exactly the same as for a minimum with one difference: to select maximum press **4** in the CALC menu.

Press **shift** > CALC **trace** to access the CALC menu. Select maximum **4** and set the left bound by tapping on the graph with your finger to place the cursor left of the maximum. Alternatively, you can enter an x-value and press **enter**. Do the same for the right bound to enclose the maximum but make sure that the bounds include one maximum of the function only.

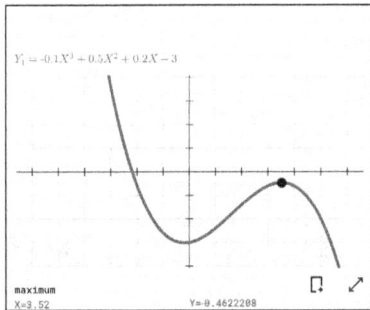

Finding a maximum (turning point) at (-3.52|-0.462) of the function Y1=-0.1X³+0.5X²+0.2X-3.

4.8 Intersection of two Functions

Start by pressing [shift] > CALC [trace] to access the CALC menu and select intersect [5]. Now the first function must be selected. Press [enter] to select it or use your finger to tap on another function, then confirm with [enter]. Do the same for the second function (if you only have stored two functions, you can press [enter] two times). Finally, make a guess near the point of intersection and press [enter].

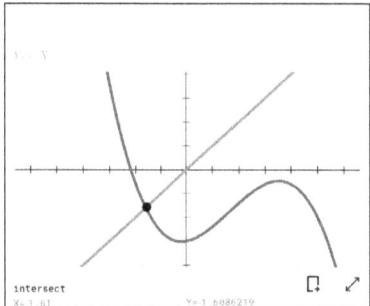

The intersection of the two functions Y1=$-0.1X^3+0.5X^2+0.2X-3$ and Y2=X is at the point (-1.61|-1.609).

4.9 Draw Derivative

You can draw the derivative by entering it as a new function in the Y= menu. Press [math] > nDeriv [8] to paste in the template. Subsequently, insert the name of the function by pressing [alpha] > F4 [draw] and choose your function. Use the arrow keys ✤ or your fingers to select the empty box behind "X=". Press [x] to enter an "X" which means that it will draw the derivative for any x-value and not just one single value. Finally, press [graph] to draw the derivative.

If you only want to know the slope at one point of the function, press [shift] > CALC [trace] to access the CALC menu. Select dy/dx [6] and use the arrow keys ◀ ▶ or your fingers to move the cursor to your desired point and press [enter]. Alternatively, you can enter an x-value and press [enter].

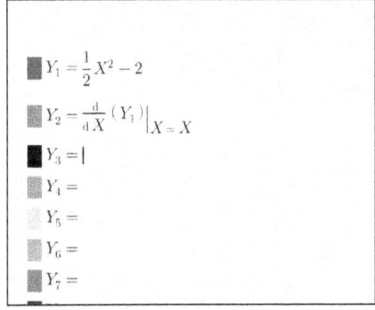

Syntax to enter the derivative as a new function Y2.

4.10 Inflection Points

The condition for inflection points is f''(x)=0. The turning points of the first derivative show the zero of the second derivative. Thus, you can determine the x-value of the inflection point by searching for the maxima or minima of f'(x). If you look at the graph of the function, you can decide whether it's a saddle point or an inflection point.

What you have to do step by step:

1. Enter the derivative of Y1 as a new function.
2. Find all turning points (minimum or maximum) of the derivative and note down the x-values of those points.
3. Find the y-value of that x-value of Y1. This will be the inflection point.

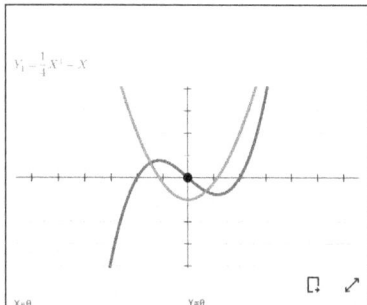

The inflection point of the function Y1=0.25X³-X, which you can easily find by looking for the turning points of the derivative of Y1.

5 SOLVE EQUATIONS

5.1 Polynomial

Insert your polynomial equation as a function in the y= menu. Make sure the polynomial equation equals 0. Press shift > CALC trace to access the CALC menu. Select zero 2 and find all zeros of the function, which are the solutions to your equation.

Read the chapter Zeros of a Function for additional information.

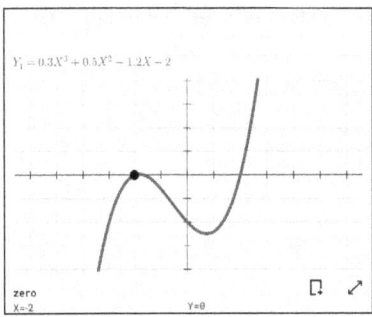

First zero (solution) at X=-2 of the graphed polynomial equation $0.3X^3+0.5X^2-1.2X-2=0$.

5.2 Solve any Equation

You can graphically solve any equation by doing the following: Access the y= menu and enter the part of the equation as Y1 which is left of the equal sign. Enter the right part of your equation as Y2. Press shift > CALC trace to access the CALC menu and select intersect 5. Then find all points of intersection, which are the solution for the equation.

Go to the chapter Intersection of two Functions to read more about how to find the intersection of two functions.

Let's take the equation $4X^4-2 = 0.5X^3+X^2$ as an example, which you want to solve. Start by entering $4X^4-2$ as Y1 and $0.5X^3+X^2$ as Y2. Then find all points of intersection. The x-coordinates are the solutions for the equation.

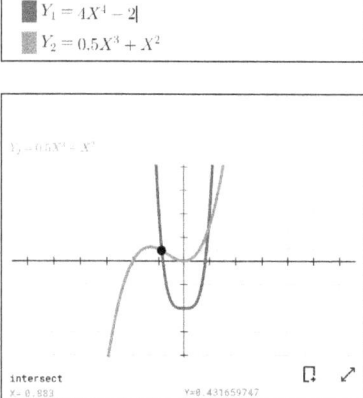

First intersection (solution) of a graphed equation at X=-0.883.

5.3 Solve Quadratic Equations

You can quickly solve quadratic equations that have the format $AX^2+BX+C=0$. Press the [solve] key and select Quadratic Equation Solver [1]. Enter the values for A, B, and C and confirm with [enter]. The app displays the solution for X1 and X2 underneath your input.

```
AX² + BX + C = 0
A = 1
B = -10
C = 16
X₁ = 8, X₂ = 2
```

6 INTEGRAL CALCULUS

6.1 Calculate Integral

Start from the home menu and press [math] > fnInt [9] to paste in the integral template. Enter the lower limit and use the right arrow key ▶ to move the cursor to the upper limit. Press ▶ again and enter your function. When you are done, press [enter] to calculate the integral.

Instead of entering a new function you can also insert "Y1" to get a function you've already stored. Press [sign] > F4 [draw] and choose your function.

$$\int_0^2 \left(-\tfrac{1}{2}X^2 + 2\right) dX$$
$$2.666666668$$
$$\int_0^2 (Y_1) dX$$
$$21.59999995$$

Calculated integral of two different functions from X=0 to X=2.

6.2 Integral in GRAPH Menu

Press `shift` > CALC `trace` to access the CALC menu and select ∫f(x)dx `7`. Your calculator shows you the graph of the function, and you can set a lower limit. Use your finger to place the cursor at your desired location and press `enter` to set the limit. Alternatively, you can enter an x-value by using the calculator keyboard and press `enter`. Do the same for the upper limit.

Attention: To calculate the area of a function, that intersects the x-axis, you have to integrate from zero to zero and sum up the integrals. Otherwise it will subtract the area under the x-axis from the area above the x-axis. There is also a way to calculate the area all at once. Read the following chapter to learn more about it.

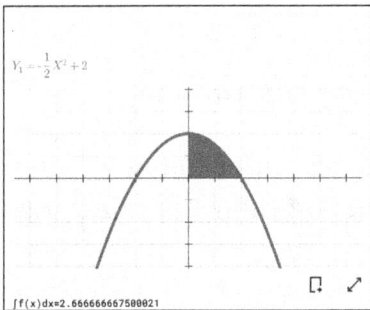

Colored area under the function and x-axis.

6.3 Find Area with Absolute Value

This method will help you to find the area between a function and the x-axis even if the function intersects the x-axis.

Enter your function as Y1 and move the cursor to the line Y2. Press `math` > ◀ to access the NUM menu and select abs `1`. Press `alpha` > F4 `draw` and choose Y1.

From here, proceed as you would to calculate a normal integral: Press `shift` > CALC `trace` to access the CALC menu and select ∫f(x)dx `7`. Your calculator shows you the graph of the function, and you have to set a lower limit. Use your finger to place the cursor at your desired location and press `enter` to set the limit. Alternatively, you can enter an x-value by using the calculator keyboard and press `enter`. Do the same for the upper limit.

The zeros of the function are the limits to calculate the area.

Important: Make sure to calculate the integral under the new function (Y2). Use your finger to choose the function and then start entering the bounds.

It is useful to deactivate the function Y1 (then it won't be drawn). To deactivate, position the cursor in the line of Y1 and press `enter`. It will be grayed out.

$$Y_1 = \frac{1}{3}X^3 - 2X^2 + \frac{8}{3}X$$
$$Y_2 = \text{abs}(Y_1)$$

Drawing Y1 with absolute value to calculate the area between the function and x-axis at once.

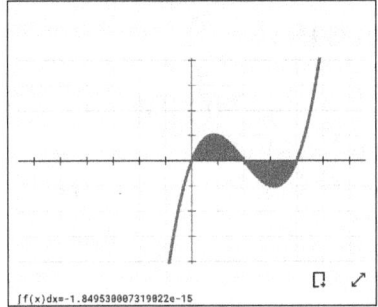 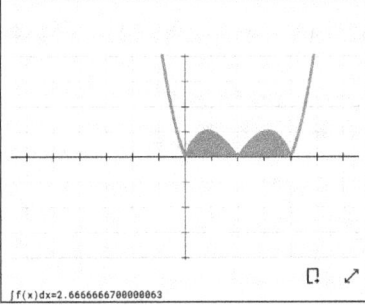

The screenshots above show the difference by calculating the integral of a normal function (left) and the same function graphed with absolute value (right).

6.4 Area between two Functions

Enter both functions Y1 and Y2 in the [y=] menu and place the cursor in the line of Y3. Press [math] > ⬇ to access the NUM menu and select abs [1]. Press [alpha] > F4 [draw] and choose Y1, press [−] and insert Y2. This will subtract both functions, so you can calculate the area between both functions.

From here, proceed as you would calculate a normal integral: Press [shift] > CALC [trace] to access the CALC menu and select ∫f(x)dx [7]. Your calculator shows you the graph of the function, and you have to set a lower limit. Use your finger to place the cursor at your desired location and press [enter] to set the limit. Alternatively, you can enter an x-value by using the calculator keyboard and press [enter]. Do the same for the upper limit.

Important: Make sure to calculate the integral under the new function (Y3). Use your finger to choose the function and then start entering the bounds.

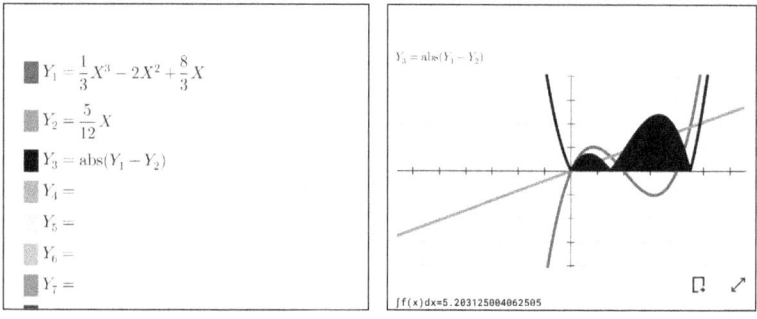

Subtract Y1-Y2 with absolute value to calculate the area between both functions.

6.5 Integral Function

The following steps will show you how to draw the integral function. First, insert your function in the [y=] menu and place the cursor in the line of a free slot. Press [math] > fnInt [9] to paste in the integral template.

You must insert one fixed limit (lower or upper) and the other limit will be variable. The y-value of the function will be the value of the integral. For instance, if you enter 0 and X, the y-value of the new function will be -8/3 at x=2, which is the same as the integral from 0 to 2.

Enter the values for your limits and use the arrow keys ✸ or your fingers to navigate through the integral template. Inside the parenthesis, insert the name of the function by pressing [alpha] > F4 [draw] and choose your function. Finally, press [graph] to draw the integral function.

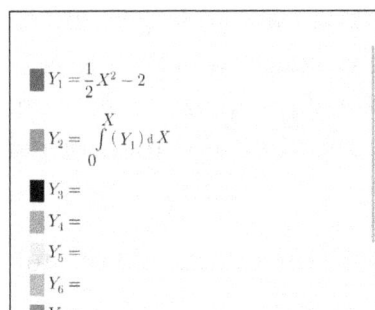

7 MATRICES

7.1 Save Matrix

This will be the first step before you can do any further calculations with a matrix.

To enter the values of your matrix, start by pressing [shift] > MATRIX [x⁻¹] to access the matrix menu. Press the right arrow key ◄ ◄ two times or click on the EDIT tab. You can choose the letter (A, B, C, etc.) under which you want to store the matrix. Select the corresponding letter and press [enter].

Next, enter the number of rows, press the right arrow key ◄ and then enter the number of columns your matrix is going to have. Press [enter]. You will see a matrix full of zeros. To enter your own values, simply override the existing matrix by entering a number and then pressing [enter] to get to the next field.

After you are done, you can exit the menu and the matrix will be stored.

```
MATRIX[A] 3 x4
2      -4      0      -6
1       2     -6       0
2       4     -6       4




[A](3,4)=-4.0
```

If you need the matrix only one time, you can enter it with the matrix shortcut menu, located under [alpha] > F3 [zoom]. Then change the number of rows and columns by tapping on the "1x1" with your finger (behind MATRIX in the new window). Fill in your matrix values and press the INSERT button.

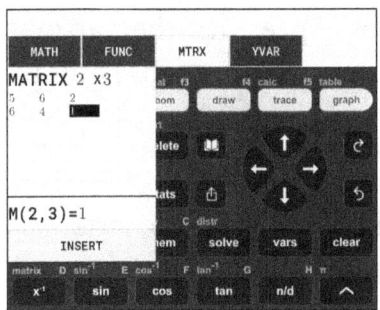

Furthermore, you can store any matrix you've created or calculated. Press [→] > [shift] > MATRIX [x⁻¹] and choose the letter under which the matrix will be stored (A, B, C, ...), then press [enter].

$$\begin{bmatrix} 5 & 6 & 7 \\ 5 & 3 & 1 \end{bmatrix} \to [B]$$
$$\begin{bmatrix} 5 & 6 & 7 \\ 5 & 3 & 1 \end{bmatrix}$$

This matrix is now stored as matrix [B] and can be used for further calculations.

7.2 Delete Matrix

Deleting a matrix is not as easy as you'd think. Therefore, it makes sense to overwrite the existing matrix if the old matrix is no longer needed. It's faster than deleting it.

To edit the matrix, go to [shift] > MATRIX [x⁻¹] and press the right arrow key ◄ ► two times to access the EDIT menu, choose your matrix and press [enter].

If you want to delete a matrix, you can do this in the MEM menu by pressing the [mem] key. Select MATRIX [4] and highlight the letter of the matrix you want to delete by using the arrow keys or your finger. Then press the [delete] key and confirm the alert box.

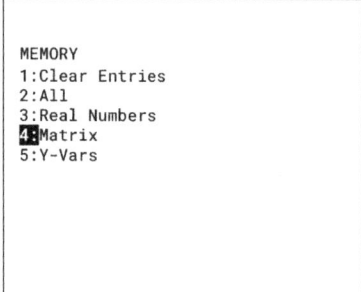

7.3 Put into Row-Echelon Form

To put your matrix into row-echelon form, paste the command to the home menu by pressing [shift] > MATRIX [x⁻¹] to access the matrix menu. Press the right arrow key ◆ or click on the MATH tab and select ref [8].

Next, enter the letter of your matrix by pressing [shift] > MATRIX [x⁻¹] to access the NAMES matrix menu. Select the corresponding letter and press [enter] to paste it to the calculator. Alternativly, Taculator has shortcuts for the matrices [A], [B], and [C] under the keys [4], [5], and [6]. Simply press [shift] and one of the buttons to directly paste the matrix.

Finally, press [enter] to execute the command which puts the matrix into row-echelon form.

$$\text{ref}([A])$$
$$\begin{bmatrix} 1 & 2 & 0 & 3 \\ 0 & 1 & -0.75 & 0.25 \\ 0 & 0 & 1 & 0.5 \end{bmatrix}$$

This command doesn't solve the matrix. In most cases, you will need to find the reduced row-echelon form of a matrix to solve it, which would be the command "rref(".

7.4 Solve Matrix (Reduced Row-Echelon Form)

To put your matrix into reduced row-echelon form, paste the command to the home menu by pressing `shift` > MATRIX `x⁻¹` to access the matrix menu. Press the right arrow key ▶ or click on the MATH tab and select rref `9`.

Next, enter the letter of your matrix by pressing `shift` > MATRIX `x⁻¹` to access the NAMES matrix menu. Select the corresponding letter and press `enter` to paste it to the calculator. Finally, press `enter` to execute the command which puts the matrix into reduced row-echelon form.

You can also use it to solve a system of linear equations. The output of the example above shows the solution x=1.75, y=0.625, and z=0.5.

$$\text{rref}([A])$$
$$\begin{bmatrix} 1 & 0 & 0 & 1.75 \\ 0 & 1 & 0 & 0.625 \\ 0 & 0 & 1 & 0.5 \end{bmatrix}$$

Keep in mind: Not all systems of linear equations have unique solutions like the example above.

1. The system has no solution if one diagonal element is equal to zero. Another condition is that the number in the right column of the same row isn't equal to zero.
2. The system has infinitely many solutions if the diagonal element and the number in the right column are equal to zero.

7.5 Transpose a Matrix

Transposing a matrix means turning all the rows of a given matrix into columns and vice versa. The matrix gets flipped along its main diagonal.

To transpose a matrix, paste the letter of your matrix by pressing shift > MATRIX x^{-1} to access the NAMES matrix menu. Select the corresponding letter and press enter. Subsequently, you need to paste the "T" command behind the letter of your matrix. Press shift > MATRIX x^{-1} to access the matrix menu. Press the right arrow key ◀ or click on the MATH tab and select T 2. Finally, press enter to execute the command which transposes your matrix.

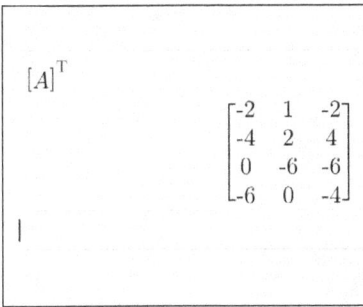

7.6 Identity Matrix

The identity matrix always has as many rows as columns. For example, enter 4 after the "identity" command to get a 4x4 identity matrix.

Start by pressing shift > MATRIX x⁻¹ to access the matrix menu. Press the right arrow key ▶ or click on the MATH tab and select identity 5 to paste the command to the calculator. Afterwards, enter the number of rows and columns you want the matrix to have and press enter to execute it.

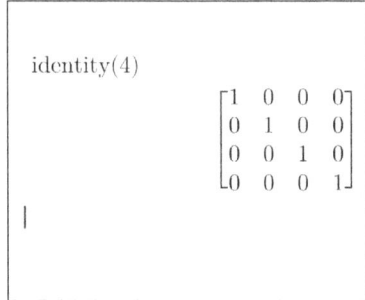

Identity matrix with four rows and columns.

7.7 Inversion of a Matrix

You are only able to calculate the inversion of a matrix with the same number of rows and columns.

To transpose a matrix, paste the letter of your matrix by pressing [shift] > MATRIX [x⁻¹] to access the NAMES matrix menu. Select the corresponding letter and press [enter]. Subsequently, press the [x⁻¹] key to paste "⁻¹" behind the letter of your matrix which will inverse it. Finally, press [enter] to execute the command.

$$[A]^{-1}$$
$$\begin{bmatrix} -0.125 & 0.25 \\ -0.1875 & -0.125 \\ -0.0833333333 & -0.1666666667 \end{bmatrix}$$
ans▶Frac
$$\begin{bmatrix} -\frac{1}{8} & \frac{1}{4} & -\frac{1}{4} \\ -\frac{3}{16} & -\frac{1}{8} & \frac{1}{8} \\ -\frac{1}{12} & -\frac{1}{6} & 0 \end{bmatrix}$$

Inversion of the matrix A. The **>Frac** command will display the answer in fraction form.

7.8 Determinant of a Matrix

You are only able to calculate the determinant of a matrix with the same number of rows and columns (it must be square).

Start by pressing **shift** > MATRIX **x⁻¹** to access the matrix menu. Press the right arrow key ▶ or click on the MATH tab and select **det** **1** to paste the command to the calculator. Afterwards, paste the letter of your matrix by pressing **shift** > MATRIX **x⁻¹** to access the NAMES matrix menu. Select the corresponding letter and press **enter**. Finally, press **enter** to calculate the determinant.

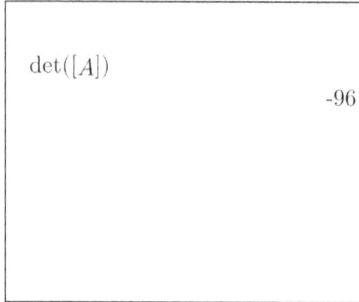

Determinant of matrix A.

7.9 Matrix Arithmetic Operations

- **Scalar Multiplication**: Enter the value of the scalar multiple and then paste the name of the matrix.

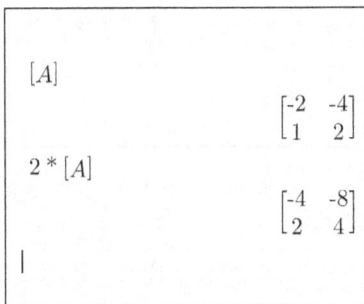

- **Addition & Subtraction**: Note that both matrices must have the same dimensions. Just combine them with + or -.

$$\begin{bmatrix} 3 & 2 & 1 \\ 4 & 5 & 6 \end{bmatrix} + \begin{bmatrix} 1 & 4 & 7 \\ 3 & 5 & 7 \end{bmatrix}$$
$$\begin{bmatrix} 4 & 6 & 8 \\ 7 & 10 & 13 \end{bmatrix}$$

- **Multiplication**: Note that the number of columns in the first matrix must equal the number of rows in the second matrix. Use * to multiply the matrices.

$$\begin{bmatrix} 3 & 2 & 1 \\ 4 & 5 & 6 \end{bmatrix} * \begin{bmatrix} 4 & 5 \\ 12 & -2 \\ 7 & 1 \end{bmatrix}$$
$$\begin{bmatrix} 43 & 12 \\ 118 & 16 \end{bmatrix}$$

- **Power of a Matrix**: The matrix must be square to find the power, and only positive integers can be used.

$$\begin{bmatrix} 1 & 2 \\ 4 & 8 \end{bmatrix}^2$$
$$\begin{bmatrix} 9 & 18 \\ 36 & 72 \end{bmatrix}$$

8 STATISTICS & PROBABILITY

8.1 Permutations, Combinations & Factorials

a. Factorials

You often need factorials to solve probability problems. To evaluate factorials, enter the number and press `math` > PROB ◆ ◆ to access the probability menu. Press `4` to enter the factorial symbol "!" and press `enter`.

As a quick reminder: 5! = 5*4*3*2*1.

```
5!
              120
5*4*3*2*1
              120
```

b. Permutation and Combination

Permutation:
You have a given a set of different items (**n**), in how many ways can you select <u>and order</u> a specific number (**r**) of them?

Example: In how many ways can a president, a treasurer and a secretary be chosen from among 7 candidates? Because each position is different, the order is important. n = 7 candidates, r = 3 positions. It is a permutation and you should use **nPr**.

Combination:
You have given a set of different items (**n**), in how many ways can you select a specific number (**r**) of them?

Example: In how many ways can 3 equivalent positions be divided among 7 candidates? Because all 3 positions are equal, the order is NOT important. n = 7 candidates, r = 3 positions. It is a combination and you should use **nCr**.

Use your calculator to solve the problem: Start on the home menu and enter the number for **n**. After that press **math** > PROB ◆ ◆ to access the probability menu. Press **nPr** 2 to enter the permutation command or press **nCr** 3 for combinations. Enter the number for **r** and press **enter**.

```
7 nPr 3
              210
7 nCr 3
               35
```

c. Binomial Theorem

If you are dealing with binomials of high degrees, it can be a pain to calculate them by hand. What's the fifth term in the binomial expansion of $(3x+2)^6$?

Use your calculator, together with this formula: $(nCr)(a)^{n-r}(b)^r$. Identifying a and b is quite easy as you should be familiar with the binomial formula $(a+b)^n$, so **a=3** and **b=2**. The power of the binomial is **n=6**. For r, you have to keep in mind that it is always one less than the number of the term you want to find, **r=4**.

Now plug the variables into the formula $(nCr)(a)^{n-r}(b)^r = (6C4)(3x)^{6-4}(2)^4$. Evaluate the formula in your calculator:

```
(6 nCr 4) * 3^(6-4) * 2^4
              2160
```

The fifth term of the binomial expansion of $(3x+2)^6$ is **$2160x^5$**.

8.2 Random Numbers

a. Random Decimals

Generating random decimal numbers is quite easy. Press **math** > PROB ◆ ◆ to access the probability menu. Press rand **1** to enter the random command and press **enter**. This will generate a number between 0 and 1. If you need a higher random number, add **10*** before the **rand** command, which generates numbers between 0 and 10.

```
rand
                    0.4554484895
rand
                    0.3035511925
10 * rand
                    2.809652384
|
```

b. Random Integers

To generate random integers, press **math** > PROB ◆ ◆ to access the probability menu. Select randInt(**5** to paste in the randInt command.

First enter the lower limit (smallest number), press the **,** key to insert a comma, then enter your upper limit (highest number) and confirm with **enter**. For example, if you set "0" as the lower limit, "0" will be the smallest integer you are able to get.

If you need more than one random integer, you can enter a value for "n" which is the number of elements. It signifies how many random integers will be generated. Insert another comma **,** after the upper limit and enter a value for "number of elements".

```
randInt(0, 20)
                                    2
randInt(0, 20)
                                   14
randInt(0, 20, 5)
                    {4, 5, 12, 0, 13}
|
```

c. Random Integers No Repetition

The calculator could generate the same integer twice. To avoid this, use the command **randIntNoRep,** which generates integers with no repetition. Press [math] > PROB ▸ ▸ to access the probability menu. Select randIntNoRep([6] to paste in the command. Enter a lower and upper limit separated by a comma [,] and press [enter].

```
randIntNoRep(0, 5)
                        {5, 1, 4, 0, 2, 3}
```

8.3 Lists & Statistical Data

a. Enter and Delete Lists

To access the STAT list editor, press [stats] > Edit... [1]. You will see the predefined lists L1 to L6 which you can use to enter your values. Use the arrow ✣ keys or your fingers to scroll through the different lists.

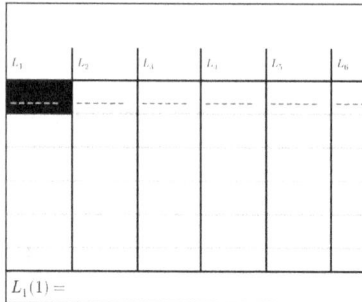

Simply enter the values and press [enter] after each value.

To delete data from a list, use the ✣ arrow keys to place the cursor on the name of the list. In the screenshot you see that the first row and column "L1" is highlighted in black.

Now press [clear] and confirm with [enter], which deletes all entries in that list. The empty list will remain.

To delete the entire list, including all data in that list, place the cursor on the name of the list and press [delete]. However, I don't recommend using the [delete] button because it is faster to clear all data in a list and refill it instead of deleting it completely and creating a new list afterward.

b. Insert a New List

Inside the STATS list editor, you can insert a new list by placing your cursor on the name of an existing list. Press [shift] > INS [delete] to insert a new list. It will appear left of the list highlighted by the cursor. Give your list a name (it is helpful to use Alpha-lock by pressing [shift] > [alpha] and then enter the name of the list) and press [enter].

c. Sorting Lists

Lists can be sorted in ascending (SortA) and descending (SortD) order.

Enter your list and go to the home screen. Press [stats] > SortA [1] to sort the list in ascending order or press SortD [2] to sort descending. Enter the list name by pressing [shift] > LIST [stats] and use the ✥ arrow keys to paste in the name of the list you want to order. Finally, press [enter].

If you are using the default list names (L1, L2, L3) you can quickly enter them by pressing [shift] > [1] to [3].

8.4 Histogram, Box Plot & Scatter Plot

a. Histogram and Box Plot

A histogram and box plot use data from one variable only (one list). To plot a histogram or box blot, enter your data in the STATS list editor first. Press [stats] > Edit [1] to access it and enter your data.

You can take list L1={5,22,15,12,13,16,23} as an example.

To construct the histogram or box plot, a plot must be turned ON. To do so, press [shift] > PLOT [y=] and press [1] to edit the first plot. Use the ✸ arrow keys or your finger to highlight ON and press [enter].

Highlight the plot type:

- 📊 for a histogram
- 📦 for a box plot

Afterwards, press ⬇ to move the cursor to the **Xlist** line where you need to enter the name of your data list. Press [shift] > LIST [stats] to see all available lists, select the one you want and press [enter]. Usually, the default value of **1** for the **frequency** is fine. Additionally, you can set a color by using the right and left arrow keys ▶ ◀ to scroll through all available colors.

STATISTICS & PROBABILITY

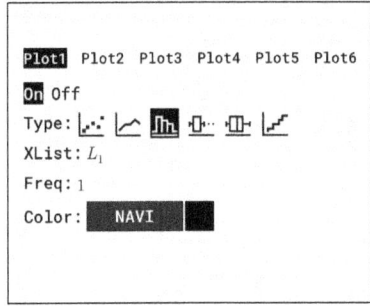

The best way to plot your data is to use the ZoomStat command, which uses an almost perfect window to display a histogram and box plot. Press `zoom` > `5` to use the zoom command and to plot your data.

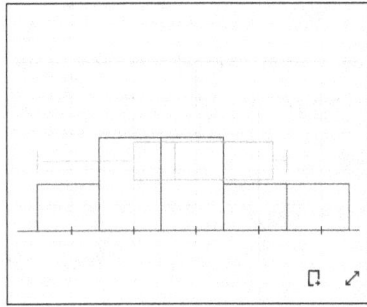

Turn off any functions entered in the `y=` menu to make sure they won't get graphed. To deselect functions, use the ✲ arrow keys to place the cursor in the line of the equation and press `enter`.

b. Adjust the Class Size

The data of the histogram above is grouped into 5 classes by your calculator. However, if you don't like how the data has been grouped, you can change the class size. To do that, go to the `window` menu and change the value of **Xscl**. A higher value means that it puts more data into one class, which will generate fewer bars.

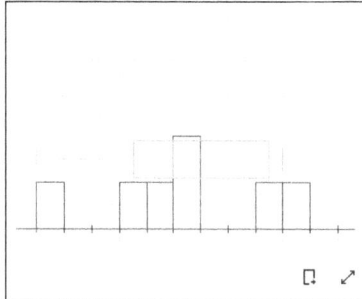

For this example, the value of Xscl is reduced from 4.5 to 2.

c. Two-Variable Data Plots

Two-Variable data plots are plots which use two lists of data. One list will be the x-coordinate and the other list will be used for the y-coordinate. The most common plots are the **scatter plot** and the **xy-line plot**.

Follow the same method described for a histogram and box plot. Make sure you enter two data lists in the STATS list editor first. Press [stats] > Edit [1] to access it and enter your data.

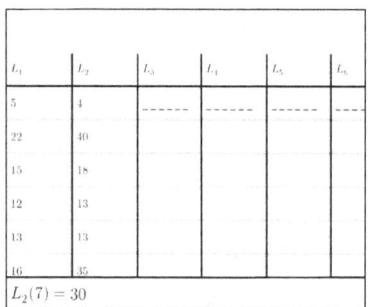

In this example, the following lists will be used:

L1 = {5,22,15,12,13,16,23}

L2 = {4,40,18,13,13,35,30}

STATISTICS & PROBABILITY

To plot the two-variable data, press [shift] > PLOT [y=] and press the number of the plot you want to edit (1,2,3, …). Use the ✪ arrow keys or your fingers to highlight ON and press [enter].

Highlight the plot type:

- ⋰ for a scatter plot
- ⊾ for an xy-line plot

Afterwards, press ⬇ to move the cursor to the **Xlist** line where you need to enter the name of your data list. Press [shift] > LIST [stats] to see all available lists, select the one you want and press [enter]. Repeat the process for the **Ylist** and choose the mark type.

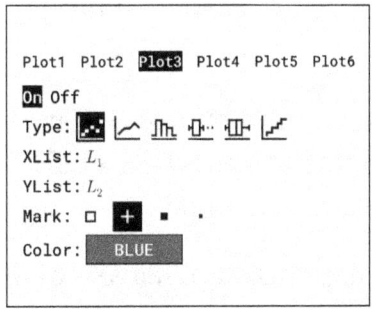

The best way to plot your data is to use the ZoomStat command, which uses an almost perfect window to display a scatter and xy-line plot. Press [zoom] > [5] to use the zoom command and to plot your data.

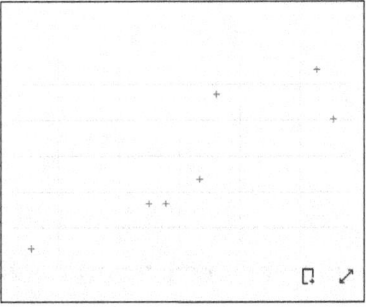

d. Tracing Plots

Any statistical data plot can be traced to see the value of the data. Press `trace` and use the arrow keys ◆ ◆ to trace the plot. If you have more than one plot, you can use the up and down arrow keys ◆ ◆ to choose between the plots. Alternatively, you can use your finger to trace.

8.5 Statistical Data Analysis

a. One-Variable Data Analysis

Enter your data in the STATS list editor first. Press `stats` > Edit `1` to access it and enter your data.

Press `stats` > ◆ to access the STATS CALC menu and choose 1-Var Stats `1`. The calculator is now asking for your data list: Press `shift` > LIST `stats` to see all available lists, select the one you want and press `enter`.

If necessary, you can provide a FreqList or skip this step (leave it blank). Press ◆ to highlight **Calculate** and press `enter`. The calculator shows you all the statistical data.

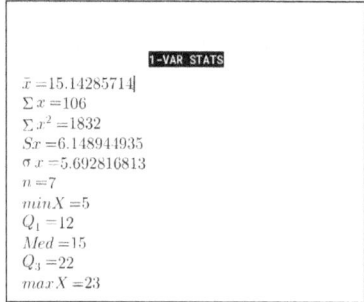

I'm using the same lists from the previous chapter in case you need an example: L1 = {5,22,15,12,13,16,23}.

The One-Variable command calculates the following statistical variables. They can also be calculated by using separate commands:

- $\bar{x}$ is the mean (average) of the elements, as returned by mean(
- Σx is the sum of the elements, as returned by sum(
- Σx² is the sum of the squares of the elements
- Sx is the sample standard deviation, as returned by stdDev(
- σx is population standard deviation
- n is the number of elements in the list, as returned by dim(
- minX is the minimum value, as returned by min(
- Q1 is the first quartile
- Med is the median, as returned by median(
- Q3 is the third quartile
- maxX is the maximum value, as returned by max(

b. Two-Variable Data Analysis

Enter two lists of data and press [stats] > ◆ to access the STATS CALC menu and choose 2-Var Stats [2]. The calculator is now asking for your **Xlist**: Press [shift] > LIST [stats] to see all available lists, select the one you want and press [enter]. Repeat the process for your **Ylist**. If necessary, you can provide a FreqList or skip this step (leave it blank). Press ◆ to highlight **Calculate** and press [enter]. The calculator shows you all the statistical data.

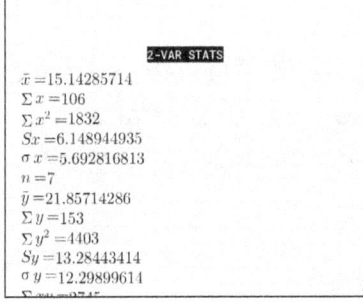

The Two-Variable command calculates the following statistical variables:

- $\bar{x}$ is the mean (average) of the first list
- Σx is the sum of the first list
- Σx^2 is the sum of the squares of the first list
- Sx is the sample standard deviation of the first list
- σx is population standard deviation of the first list
- n is the number of elements in both lists
- y is the mean (average) of the second list
- Σy is the sum of the second list
- Σy^2 is the sum of the squares of the second list
- Sy is the sample standard deviation of the second list
- σy is population standard deviation of the second list
- Σxy is the sum of products of each matching pair of elements in the lists
- minX is the minimum element of the first list
- maxX is the maximum element of the first list
- minY is the minimum element of the second list
- maxY is the maximum element of the second list

c. More Statistics Commands

Some more statistics commands are located inside the LIST MATH menu. However, you get the data for most commands by using the **1-Var Stats** command. If you only need one value, it makes sense to use the commands of the LIST MATH menu.

Press [shift] > LIST [stats] > ◄ ◄ to access the LIST MATH menu. Choose a command: for example, press variance [8] to paste the command to the home screen. Press [shift] > LIST [stats] to see all available lists, select the one you want to use and press [enter].

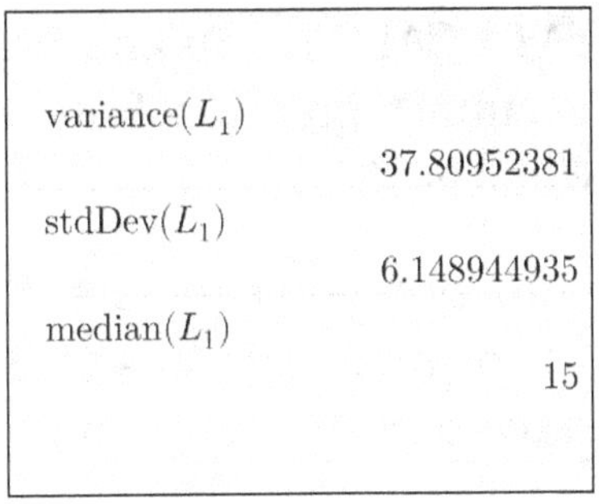

```
variance(L₁)
                    37.80952381
stdDev(L₁)
                    6.148944935
median(L₁)
                              15
```

8.6 Regression

Let's assume you have four points and want to find the equation of a cubic function.

- f(-5) = 0
- f(-2.5) = 0.5
- f(0) = 0
- f(2.5) = -0.5

The value inside the parentheses is the x-value. At x = -5 the function has a y-value of 0.

To find the equation of the function through regression, enter all x-values in one list and all y-values in another list. Press **stats** > Edit... **1** to open the list editor and enter the values like on the screenshot below.

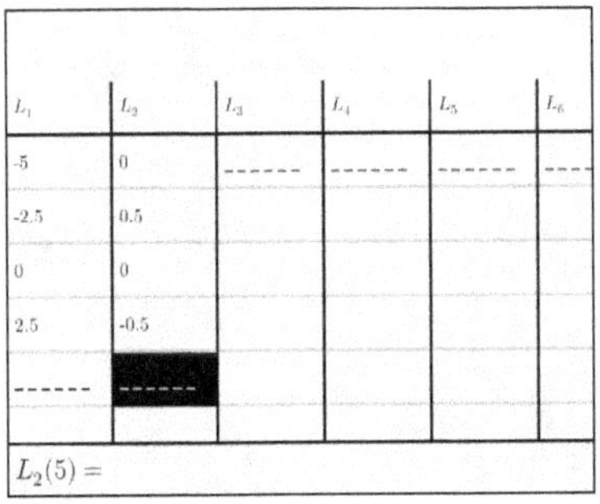

L_1	L_2	L_3	L_4	L_5	L_6
-5	0	---	---	---	---
-2.5	0.5				
0	0				
2.5	-0.5				
---	---				

$L_2(5) =$

Afterwards, press [stats] > ▶ to access the STATS CALC menu and press CubicReg [5] to open the input menu for the cubic regression. Enter the Xlist which is L1 in this example and Ylist which is L2.

If you want to store the equation as a new function you can enter it in the field "Store RegEQ". For example, use "Y1" by pressing [alpha] > F4 [draw] to paste the Y1 function which will be used to store the new equation. Note that if there is a function stored as Y1 it will be overwritten.

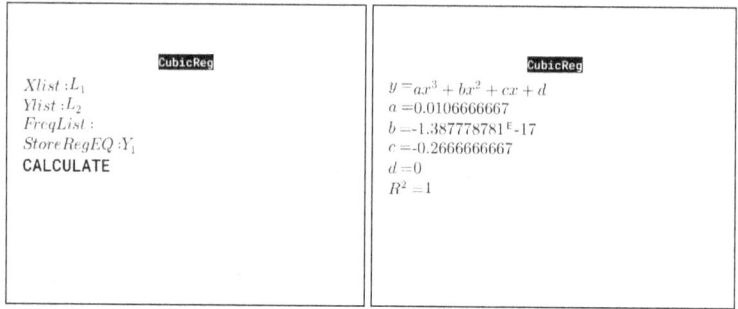

There are different types of regressions, choose...

- **LinReg** for lines
- **QuadReg** for parabolas
- **CubicReg** for functions of degree 3
- **QuartReg** for functions of degree 4
- **LnReg** for logarithmic functions
- **ExpReg** for exponential functions
- **PwrReg** to fit a power regression model
- **Logistic** to fit a logistic regression model
- **SinReg** for trigonometric functions

www.ingramcontent.com/pod-product-compliance
Lightning Source LLC
Chambersburg PA
CBHW080520220526
45465CB00006B/2551